SpringerBriefs in Electrical and Computer Engineering

For further volumes:
http://www.springer.com/series/10059

Ana M. Barbancho • Isabel Barbancho
Lorenzo J. Tardón • Emilio Molina

Database of Piano Chords

An Engineering View of Harmony

 Springer

Ana M. Barbancho
University of Málaga
Málaga, Spain

Isabel Barbancho
University of Málaga
Málaga, Spain

Lorenzo J. Tardón
University of Málaga
Málaga, Spain

Emilio Molina
University of Málaga
Málaga, Spain

Additional material to this book can be downloaded from http://extras.springer.com

ISSN 2191-8112   ISSN 2191-8120 (electronic)
ISBN 978-1-4614-7475-3   ISBN 978-1-4614-7476-0 (eBook)
DOI 10.1007/978-1-4614-7476-0
Springer New York Heidelberg Dordrecht London

Library of Congress Control Number: 2013937729

Printed on acid-free paper

Springer is part of Springer Science+Business Media (www.springer.com)

To our families

Preface

Transcription of music can be defined as the process of analysing an acoustic musical signal with the aim of writing down the musical parameters of the sound. Automatic transcription of music has been a subject of increasing interest during the last decades. This task can be faced dividing the targeted musical signals into two main categories: monophonic musical signals, leading to the transcription of music in which single notes sound, and polyphonic musical signals, related to the transcription of music in which several notes sound simultaneously.

One of the first main problems that a researcher in automatic music transcription has to face is that of the availability of an appropriate music database. The database must contain a sufficiently large number of samples to train and test a transcription system. Also, note that the database should provide with the correct annotation of each sample.

Musical instrument databases commonly used only contain individual notes of the instruments considered. This issue easily constitutes a serious drawback for piano-related activities since it is a polyphonic instrument in which the transcription of polyphonic sounds plays an important role.

The purpose of this book is to present a database of piano chords that can be used in research tasks in the context of polyphonic piano transcription.

When we originally conceived this project in the *Universidad de Málaga* (UMA) in 2007, the objective was simple: to design a database of piano chords that could be used in our research activities on polyphonic piano transcription. As time passed, the concept of the database grew larger and larger, as well as the interest in the international community for this sort of resources. Finally in 2011, we decided to look for a publisher for our work. In the IEEE ICASSP2011 Conference in Prague, we finally found that a highly prestigious publisher, Springer, was interested in our work.

Now, we have developed the most complete database of piano chords that exist nowadays and a book that supports it.

This book is mainly intended for researchers and graduate students in automatic music analysis and transcription systems. Sometimes, the knowledge of the researches on Western harmony is limited and the books of harmony are strongly oriented to musicians. Trying to fill this gap, this book contains an engineering view of harmony. The descriptions included should help researchers to understand the foundations of Western harmony and to comprehend the content of the presented UMA-Piano chord database. Thus, Chap. 2 contains the foundations of harmony. This chapter is self-contained and it can be used to study Western harmony independently of the database developed. Afterwards, Chap. 3 provides a detailed description of the database. This description is completed with the help of Appendix A in which all the types of recorded chords are summarised including musical examples. Finally, Chap. 4 presents the summary and a discussion about the database developed and the possibility of future improvements.

The piano chords database can be found at http://Extras.Springer.com.

The authors are grateful to the many people who have helped both directly and indirectly to write this book and to compile the complete database presented. The database is expanding and the authors will thank collaborations that help to grow and improve the database.

This work has been funded by the Ministerio de Economía y Competitividad of the Spanish Government under Project No. TIN2010-21089-C03-02 and Project No. IPT-2011-0885-430000 and by the Ministerio de Industria, Turismo y Comercio, under Project No. TSI-090100-2011-25.

Málaga, Spain   Ana M. Barbancho
                                                                  Isabel Barbancho
                                                                 Lorenzo J. Tardón
                                                                   Emilio Molina

Contents

List of Figures

List of Tables

Chapter 1
Introduction

Abstract This book describes the content and the specifications of the UMA-Piano chord database, a piano database compiled specifically for research purposes. Shared database in the field of music information processing are now becoming common and allow to compare different methods with the same test set. There exist several examples of instrument sound databases but most of them contain just single notes or include a very limited number of chords and few types of chords. The presented UMA-Piano chord database includes 275,040 chords with polyphony number ranging from 1 to 10, each of which recorded with three dynamics levels and three different playing styles. The recorded chords have musical meaning as well as different difficulty levels of detection.

1.1 Interest of the UMA-Piano Chord Database

The polyphonic transcription of piano chords is an interesting signal processing problems that can be faced in different ways [1–6] and [7]. All these transcription systems have to solve two basic problems: to determine the possibly played notes and to determine how many notes were played [8]. Always, a significant number of samples is necessary to design, tune and evaluate the systems.

The databases that are commonly used are RWC-Musical Instrument Sound database [9] and McGill University Master Samples [11]. The problem of theses databases is that they contain only individual notes of the instruments. This is not a problem in monophonic instruments like flute, violin, etc. but it is a problem for polyphonic instruments like the piano in which the transcription of polyphonic sounds is an important task.

In the annual MIREX evaluation campaign [15], one of the proposed tasks is the evaluation of chord estimation schemes. From 2008, this task consists on determining for each frame, the played chord. Therefore, it is important to have a chord database with different levels of polyphony and difficulty of detection.

A.M. Barbancho et al., *Database of Piano Chords: An Engineering View of Harmony*,
SpringerBriefs in Electrical and Computer Engineering,
DOI 10.1007/978-1-4614-7476-0_1, © The Author(s) 2013

Also, we must be aware that the metrics are changing, in some contexts, in order to provide more musically precise and relevant evaluations as well as to encourage future research to produce more musically relevant transcriptions. These new metrics should include more detailed frame-based evaluations as several levels of chord detail (v.gr. root, triad, first extended note, second extended note) and bass note [15].

In this book, a piano chord music database is presented. This database contains chords with polyphony number ranging from 2 to 10, with three dynamics levels (*Forte, Mezzo, Piano*) and three different playing styles (*Normal, Staccato, Pedal*). It also includes the individual recording of each of the 88 piano keys with the same dynamics levels and playing styles. The recorded chords cover the full range of the piano. The total number of recorded files is 275,040, so it is a large database that contains a very significant number of chords of all the types.

The large number of chords makes the UMA-Piano chord database useful for transcription systems that need a training set of data to train the systems and a test set [26, 27].

1.2 Outline

This book is divided into four chapters. In Chap. 1, the importance of having a database of chords has been presented.

Chapter 2, presents the foundations of harmony which are needed to understand the basis of musical chords. This chapter starts with a brief description of how a musical note could be characterized. Next, the physical basis of scales and musical intervals, as well as the physical basis of occidental chord are presented.

In Chap. 3, the content of the UMA-Piano chord database is presented in detail. Finally, Chap. 4 presents the summary and a discussion of the database.

The book also includes an appendix in which the lists of the types of the recorded chords are shown including musical examples.

Chapter 2
Foundations of Harmony

Abstract This chapter presents the foundations of harmony needed to understand the basis of musical chords. This chapter starts with a brief description of how a musical note could be characterized in the time domain and in the frequency domain. Chords are based on the relation between notes. The relation between notes are the musical intervals and the musical intervals are fixed by the musical scales, so, this chapter contains a description of the physical basis of scale and musical intervals. The scales that are going to be presented here correspond to the just intonation scale, that is the most consonant scale that exists, and the equal tempered scale that is the scale actually used. The MIDI pitch number is also presented due to the usefulness to represent intervals and relations between notes. Finally, the physical basis of the occidental chords and a discussion about whether it is easy or difficult to transcribe the different chords is presented.

Harmony is a term which denotes the formation and relationships of simultaneous notes, called chords, that is, harmony refers to the combination of simultaneous pitches [16].

In order to perform musical transcription of polyphonic music (or chords), it is necessary to know how the musical notes that compose the chords can be characterized (Sect. 2.1).

Next, it is necessary to study the relations that exist between the notes in occidental music theory (Sect. 2.2). These relations are based on the occidental scales and the musical intervals that these scales determine. Among the different scales that exist in occidental music, in this chapter the just intonation scale and the equal tempered scale are presented. The just intonation or pure intonation scale can be considered as the best scale that exists because is based on the harmonic scale, that is the consistent use of harmonic intervals tuned so pure that they do not beat [10]. This scale couldn't be used in practice because since the instrument should be returned each time the key changes. On the other hand, the equal tempered scale is not perfectly tuned, but it is the scale normally used [17].

A.M. Barbancho et al., *Database of Piano Chords: An Engineering View of Harmony*,   3
SpringerBriefs in Electrical and Computer Engineering,
DOI 10.1007/978-1-4614-7476-0_2, © The Author(s) 2013

Once the relation between notes is known, the physical basis of occidental chords is justified (Sect. 2.3) This physical basis aids the musical transcription and gives rise to the idea of the difficulty of the transcription of the different chords.

Also note that the MIDI pitch number is often used to identify the frequency of the notes because it comes from a transformation of the frequency into a linear scale and so it is well known and easy to use.

2.1   Characterization of Musical Notes

Notes are the foundation of music because every classical composition can be divided into notes. Nowadays, there exist many works that study how to characterize the musical notes, derived from the importance of automatic music transcription.

The characterization of musical notes can be divided into two main block: Time domain characterization and Frequency domain characterization.

2.1.1   Time Domain Characterization of Musical Notes

In the time domain, one of the features that can be studied for a musical note is the envelope [21]. The envelope is the evolution over time of the amplitude of a sound. It is one of the most important attributes and it can be used together with other features, to identify different musical instruments of different playing techniques.

In Fig. 2.1, the envelope of the piano notes $C1$, $C4$ and $C8$ played normal and mezzo forte is shown. In this figure, it can be observed that the envelopes are quite similar, because all the notes come from a piano played with the same style and dynamic.

It is interesting to analyze the waveform of the musical notes besides the envelope. The waveform varies from one instrument to another. In the case of instruments with definite pitch, like the piano, the waveform resembles periodic signals (Fig. 2.2).

2.1.2   Frequency Domain Characterization of Musical Notes

Taking into account Fourier theory, periodic signal can be represented by a summation of sinusoids, each with a particular amplitude and phase [25] and similarly musical notes. The specification of the strengths of each of the sinusoids, usually in the form of a graph, is referred to as spectrum, or frequency domain representation of a signal. When the frequency of the sinusoids are multiple of a certain fundamental frequency, then those are called harmonics, in other cases they

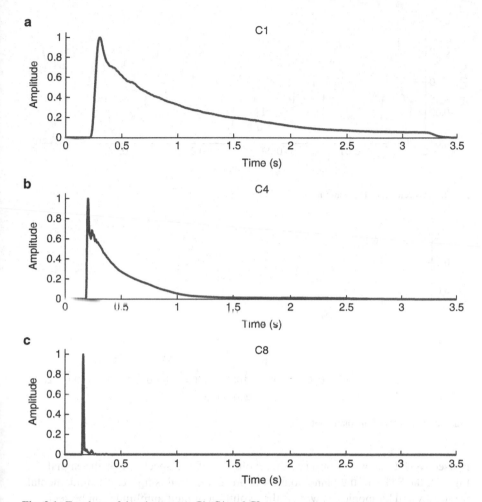

Fig. 2.1 Envelope of the piano notes $C1$, $C4$ and $C8$

are called partials. Generally, harmonics and partials can be named as overtones [23]. In the case of instruments with definite pitch, like the piano, the overtones are normally harmonics. It is noteworthy that the amplitude distribution of harmonics or overtones and the number of them can also be used to identify the notes of a certain instrument.

In Fig. 2.3, the spectrum of the piano note $C4$ is represented. In this figure the fundamental frequency and the harmonics are outlined.

The spectrum of a musical note is very useful for different purposes, such as pitch estimation. But, in other cases, it is interesting to obtain, no only information about the frequency location and amplitudes of the fundamental harmonic and partials, but also about their temporal evolution. Then, the short-time Fourier transform (STFT)

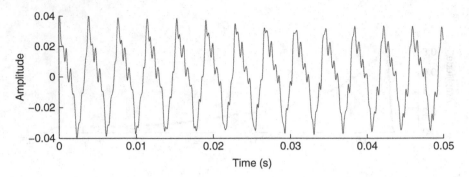

Fig. 2.2 Waveform of the piano note *C*4

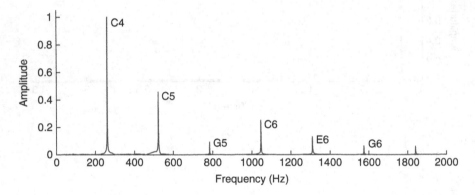

Fig. 2.3 Spectrum of the piano note *C*4

is used. STFT allows to observe the evolution of the spectrum of the signal. In Fig. 2.4, the STFT of the piano note *C*4 is presented. In this figure, the fundamental frequency and harmonics as well as the evolution of their amplitude can be seen.

2.2   Physical Basis of Scales and Musical Intervals

Before proceeding to describe the physical basis of scales and musical instruments, several definitions and concepts are presented.

A scale is a succession of notes arranged in ascending or descending order. Scales are a topic that has fascinated mathematicians as well as musicians since the time of the Greeks.

A musical interval can be defined as a ratio between two pitches and, in terms of frequency, it is described by the ratio between the respective frequencies of the involved notes. Therefore, in order to go up or down a musical interval, the

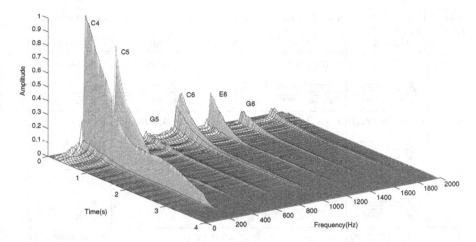

Fig. 2.4 STFT of the piano note *C*4

Fig. 2.5 Western note names
in a piano keyboard (only one
octave is labeled)

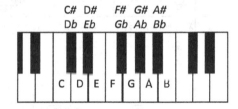

frequencies must be multiplied or divided by the corresponding number. This is due
to the fact that the frequencies in the scale follow a logarithmic relationship instead
of a linear one.

Musical intervals are defined by the musical scale used. So, the physical basis of
scales are the physical basis of the intervals.

A musical note can be identified using a letter and an octave number. On the
standard keyboard in Fig. 2.5, successive keys produce tones that are a semitone
apart. The white keys are labeled with letters A–G, and the black keys are denoted by
a letter followed by # (sharp) or *b* (flat). The 12 notes in each octave are called pitch
classes. There are 12 pitch classes, but 7 note names (C,D,E,F,G,A,B). Each note
name is separated by one tone except F from E and C from B, which are separated
a semitone interval. This is because Western modern music theory is based on the
diatonic scale. The standard reference frequency is the note *A*4 with a fundamental
frequency of 440 Hz.

Table 2.1 shows the most widely used names for the intervals between notes, the
difference in tones and semitones that correspond to the intervals, the difference in
MIDI number and an example of each interval. In Western music theory the number
and quality of an interval are typically need to name it. The number of an interval is
the number (second, third, forth, fifth, etc.) of staff positions it encompasses (v.gr.
Interval of fifth, *C*4 to *G*4) and the quality of an interval (perfect, major, minor,
augmented, and diminished) refers to the consonance of the interval (v.gr. Interval
of perfect fifth, *C*4 to *G*4).

Table 2.1 The most conventional names for intervals between notes (blank cells are included for non-existent intervals)

Interval	Diminished	Minor	Perfect	Major	Augmented
2	$X : (2bb) \; MIDI : (0)$	$X : (2b) \; MIDI : (1)$		$X : (2) \; MIDI : (2)$	$X : (2\#) \; MIDI : (3)$
3	$X : (3bb) \; MIDI : (2)$	$X : (3b) \; MIDI : (3)$		$X : (3) \; MIDI : (4)$	$X : (3\#) \; MIDI : (5)$
4	$X : (4b) \; MIDI : (4)$		$X : (4) \; MIDI : (5)$		$X : (4\#) \; MIDI : (6)$
5	$X : (5b) \; MIDI : (6)$		$X : (5) \; MIDI : (7)$		$X : (5\#) \; MIDI : (8)$
6	$X : (6bb) \; MIDI : (7)$	$X : (6b) \; MIDI : (8)$		$X : (6) \; MIDI : (9)$	$X : (6\#) \; MIDI : (10)$
7	$X : (7bb) \; MIDI : (9)$	$X : (7b) \; MIDI : (10)$		$X : (7) \; MIDI : (11)$	$X : (7\#) \; MIDI : (12)$
8	$X : (8b) \; MIDI : (11)$		$X : (8) \; MIDI : (12)$		$X : (8\#) \; MIDI : (13)$

In tonal music, a scale is an ordered set of notes typically used in a tonality (also referred to as key). The tonality is the harmonic center of gravity of a musical excerpt. Intervals in the major and minor scales are consonant intervals relative to the tonic. The diatonic scale is a seven-note musical scale comprising five whole steps and two half steps, with the pattern repeating at the octave. The major scale is a diatonic scale in which the pattern of intervals between the notes in semitones is 2-2-1-2-2-2-1, starting from a root note. The natural minor scale has the following pattern of intervals 2-1-2-2-1-2-2. A musical excerpt can be arranged in a major or minor key. Major and minor keys that share the same number of alterations (key signature) are called relative keys [13].

Because of the logarithmic nature or the frequencies, which makes the operation with intervals somehow difficult, the MIDI (Musical Instrument Digital Interface) pitch number is generally used.

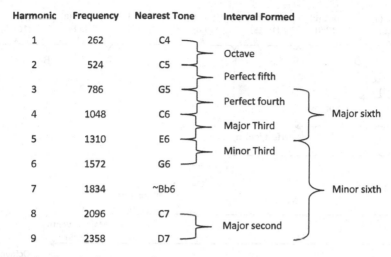

Fig. 2.6 Harmonic series for the note C4 and the intervals formed in this series

On the other hand, there are many ways to establish the frequency of each musical note or interval. In this section, we present the just intonation scale, the equal tempered scale and the MIDI pitch number.

2.2.1 Just Intonation Scale

The basis of this scale consists on selecting sounds of the diatonic scale so that the intervals of each sound with the tonic are taken from the harmonic series. The frequencies of the harmonic series are integer multiples of the fundamental frequency, therefore, these frequencies are naturally related to each other by integer ratios. In Fig. 2.6, the harmonic series for the note C4 and the intervals formed in this series are presented. It is noteworthy that the intervals are ratios of integer numbers. These intervals are the ones selected for the just intonation scale.

The intervals for the just intonation scale in C-major, are represented in Fig. 2.7.

This scale has perfect tuning due to the selection of the sound from the harmonic series and contains the most consonant intervals of music. These intervals, in descending order of consonance, are presented in the Table 2.2

It should be noted than the octave is the most consonant interval. The principle of octave equivalence, well known to musicians, is based on the fact that the octave comprises only the even harmonics, so a given tone with many harmonics already contains the octave as a subset of its partials.

Although this scale is the most consonant one, it has never been of much practical use due to the existence of two different tone intervals and three different semitone intervals. In this scale, enharmonic sound does not exist, that is, C# presents different frequency than Db. This implies that 72 sounds per octave are needed to play musical pieces in every key. So, this scale can't be used in every type of instrument.

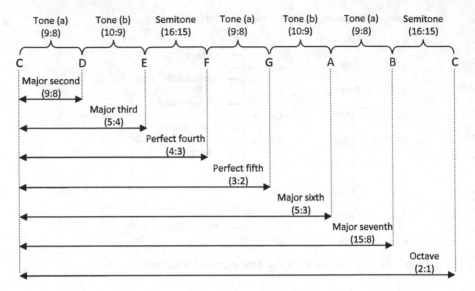

Fig. 2.7 Intervals for the just intonation scale in C-major

Table 2.2 The most consonant intervals of music in descending order of consonance

Frequency ratio	Interval	Note sample
2:1	Octave	(*C*/*C*)
3:2	Perfect fifth	(*G*/*C*)
4:3	Perfect fourth	(*F*/*C*)
5:3	Major sixth	(*A*/*C*)
5:4	Major third	(*E*/*C*)
8:5	Minor sixth	(*Ab*/*C*)
6:5	Minor third	(*Eb*/*C*)
9:8	Major second	(*D*/*C*)
15:8	Major seventh	(*B*/*C*)
16:15	Minor second	(*Db*/*C*)
16:9	Minor seventh	(*Bb*/*C*)

Therefore, it is necessary to find another scale that can be used in any instrument and make enharmonic sound possible.

2.2.2 Equal Tempered Scale

The basis of the equal tempered scale consists on dividing the octave into 12 logarithmically equal parts, or semitones. One tone is defined as a two semitones intervals. This is the only scale that fulfills the enharmonic equivalences. The intervals for the equal tempered scale in C-major are represented in Fig. 2.8.

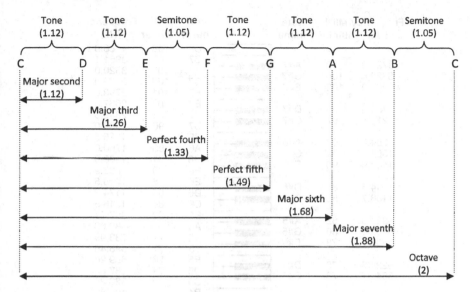

Fig. 2.8 Intervals for the equal tempered scale in C-major

Table 2.3 Comparison between the musical intervals in the just intonation scale and the equal tempered scale

Scale Interval	Just intonation	Equal tempered
Octave	2:1 = 2.00	2.00
Perfect fifth	3:2 = 1.50	1.49
Perfect fourth	4:3 = 1.33	1.33
Major sixth	5:3 = 1.66	1.68
Major third	5:4 = 1.25	1.26
Minor sixth	8:5 = 1.60	1.58
Minor third	6:5 = 1.20	1.18
Major second	9:8 = 1.12	1.12
Major seventh	15:8 = 1.87	1.88
Minor second	16:15 = 1.06	1.05
Minor seventh	16:9 = 1.77	1.78

In this scale, the octave interval is the only interval that matches up with the just intonation scale. In Table 2.3, a comparison between the musical intervals in the just intonation scale and the equal tempered scale is presented.

2.2.3 MIDI Pitch Number

MIDI is a specification for a communications protocol principally used to control electronic musical instruments. It consists of two parts: a set of commands, or the MIDI 'language', and the details for the electrical conditions in which those commands are transmitted and received [24].

Frequency (Hz)	MIDI number	Note name
3729.3	106	A#7
3322.4	104	G#7
2960.0	102	F#7
2489.0	99	D#7
2217.5	97	C#7
1864.7	94	A#6
1661.2	92	G#6
1480.0	90	F#6
1244.5	87	D#6
1108.7	85	C#6
932.33	82	A#5
830.61	80	G#5
739.99	78	F#5
622.25	75	D#5
554.37	73	C#5
466.16	70	A#4
415.30	68	G#4
369.99	66	F#4
311.13	63	D#4
277.18	61	C#4
233.08	58	A#3
207.65	56	G#3
185.00	54	F#3
155.56	51	D#3
138.59	49	C#3
116.54	46	A#2
103.83	44	G#2
92.499	42	F#2
77.782	39	D#2
69.296	37	C#2
58.270	34	A#1
51.913	32	G#1
46.249	30	F#1
38.891	27	D#1
34.648	25	C#1
29.135	22	A#0

Note name	MIDI number	Frequency (Hz)
C8	108	4186.0
B7	107	3951.1
A7	105	3520.0
G7	103	3136.0
F7	101	2793.8
E7	100	2637.0
D7	98	2349.3
C7	96	2093.0
B6	95	1975.5
A6	93	1760.0
G6	91	1568.0
F6	89	1396.9
E6	88	1318.5
D6	86	1174.7
C6	84	1046.5
B5	83	987.77
A5	81	880.00
G5	79	783.99
F5	77	698.46
E5	76	659.26
D5	74	587.33
C5	72	523.25
B4	71	493.88
A4	**69**	**440.00**
G4	67	392.00
F4	65	349.23
E4	64	329.63
D4	62	293.67
C4	60	261.6
B3	59	246.94
A3	57	220.00
G3	55	196.00
F3	53	174.61
E3	52	164.81
D3	50	146.83
C3	48	130.81
B2	47	123.47
A2	45	110.00
G2	43	97.999
F2	41	87.307
E2	40	82.407
D2	38	73.416
C2	36	65.406
B1	35	61.735
A1	33	55.000
G1	31	48.999
F1	29	43.654
E1	28	41.203
D1	26	36.708
C1	24	32.703
B0	23	30.868
A0	21	27.500

Fig. 2.9 Equal tempered frequency, note name and MIDI note number

In MIDI, the pitch of a note is encoded using a number (see Fig. 2.9). A frequency f can be easily converted into a MIDI pitch number f_{MIDI} as follows:

$$f_{MIDI} = 69 + 12 log_2(f/440) \tag{2.1}$$

Table 2.4 MIDI difference of each interval and a note sample in descending order of consonance

Interval	MIDI difference	Note sample
Octave	12	(*C* to *C*)
Perfect fifth	7	(*C* to *G*)
Perfect fourth	5	(*C* to *F*)
Major third	4	(*C* to *E*)
Major sixth	9	(*C* to *A*)
Minor third	3	(*C* to *Eb*)
Minor sixth	8	(*C* to *Ab*)
Major second	2	(*C* to *D*)
Major seventh	11	(*C* to *B*)
Minor second	1	(*C* to *Db*)
Minor seventh	10	(*C* to *Bb*)
Diminished third	2	(*C* to *Dbb*)
Diminished fifth	6	(*C* to *Gb*)
Diminished seventh	9	(*C* to *Bbb*)
Diminished octave	11	(*C* to *Cb*)
Augmented second	3	(*C* to *D#*)
Diminished sixth	7	(*C* to *Abb*)
Augmented fourth	6	(*C* to *F#*)
Augmented fifth	8	(*C* to *G#*)
Augmented octave	13	(*C* to *C#*)
Augmented third	5	(*C* to *E#*)
Augmented sixth	10	(*C* to *A#*)
Diminished fourth	10	(*C* to *Fb*)

The MIDI pitch number is based on the equal tempered scale. The relation could be seen as if the MIDI frequencies were the logarithm of the frequencies of the equal tempered scale, so that a semitone would be seen as MIDI difference of 1 and a tone as a MIDI difference of 2 and so on. In Table 2.4, the most relevant intervals in MIDI difference are presented.

2.3   Physical Basis of Occidental Chords

Chords are referred to as the simultaneous sounding of three or more tones [13]. Chords can be classified as consonant or dissonant on the basis of the following definition [13]:

- A consonant chord is a chord in which only consonant intervals (octave, perfect fifth, perfect fourth, third, sixth) are found.
- A dissonant chord is a chord that includes at least one dissonant interval (second, seventh, diminished fifth, etc.)

The elemental chord in harmony is the triad, which is a three note chord with a root, a third degree (major or minor third above the root), and a fifth degree

Fig. 2.10 *C* key: (**a**) Perfect major chord, first inversion (6) and second inversion $\binom{6}{4}$. (**b**) Perfect minor chord first inversion (6) and second inversion $\binom{b6}{4}$

(major or minor third above the third). This group of elementary chords contains the most consonant chords that exists:

- The perfect major chord, that is composed by a root, a third degree major above the root and a perfect fifth degree above the root.
- The perfect minor chord, that is composed by a root, a third degree minor above the root and a perfect fifth degree above the root.

These perfect major and perfect minor chords and their inversions form the group of consonant chords. In Fig. 2.10, an example of the perfect major chord, the perfect minor chord and their inversions for *C* key are presented.

The group of dissonant chords is wider than the group of consonant chords and contains chords with three or more different notes. The question that arises is why these chords or intervals are more consonant than others. Helmholtz [22] explained consonance by referring to Ohm's acoustical law [23], which stated that the ear performs a Fourier analysis of sound, separating a complex sound into its various partials. Helmholtz concluded that dissonance occurs when partials of the two tones produce 30–40 beats per second. The more the partials of one tone coincide with the partials of the other, the lesser the chance that beats in this range produce dissonance.

The consonance or dissonance in chords could be shown in the comparison between the spectrum of the chords and the spectrum of the individual notes that compose the chords. During the comparison of these spectra, the coincidence or not of harmonics is revealed. If harmonics coincide or are very far from each other, the result is consonant, on the other hand it is dissonant.

The most consonant chord that exist, and the most typical Western music chord, is the perfect major chord (Fig. 2.11a). It is composed by a root (v.gr. *C*4), its major third degree (v.gr. *E*6) and its perfect fifth degree (v.gr. *G*5). If we compare the spectrum of *C*4 (Fig. 2.11b) with the spectrum of *G*5 (Fig. 2.11c) and *E*6 (Fig. 2.11d), it can be seen that there is a strong harmonic relation between these notes. Hence, this chord is difficult to transcribe because of the harmonic relation between the notes that can make the transcription system consider a lower number of notes (v.gr. *C*4 and *E*6) or notes in different octaves (v.gr. *C*4, *E*5 and *G*4). Therefore, the perfect major chord, in all the polyphony levels and with different

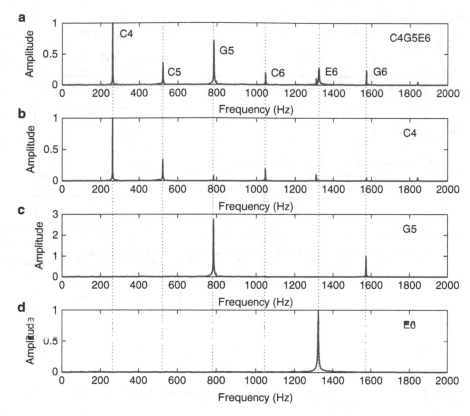

Fig. 2.11 Comparison of the spectrum of (**a**) the perfect major chord *C4E6G5* and its individual notes (**b**) *C4*, (**c**) *G5* y (**d**) *E6*

frequency separations and order of the notes as well as different doubled notes (v.gr. for polyphony number 4, we can double the root note, the third degree note and the fifth degree note) is considered difficult to detect.

Note that the relative position of the notes or frequency separation between the notes, is also important. For example, consider a perfect major chord like *C1E4G7*, this chord is easier to detect than *C4E4G4*, because in the first chord the notes are more separated in frequency, than in the second chord.

Another important aspect is the octave in which the chord is recorded. For example, the detection of the piano chord *C1E1G1* has to face the problem derived from the fact that the fundamental frequency of its notes is very small and sometimes it does not exist. On the other hand, in the piano chord *C7E7G7* the harmonics of the notes have very high frequency and sometimes they are not considered in the spectrum [6].

There exist other groups of chords common in Western music, these are less consonant chords than the previous one such as the diminished minor chord (Fig. 2.12a), among others. This chord is composed by a root (v.gr. *C4*), its minor

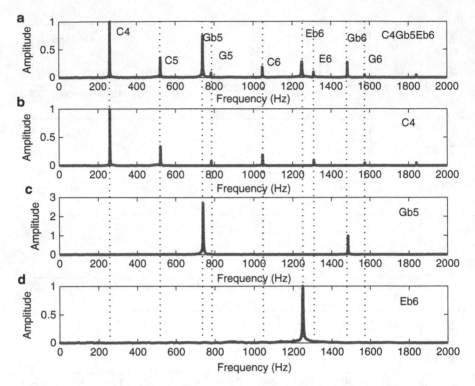

Fig. 2.12 Comparison of the spectrum of (**a**) the perfect major chord *C4Eb6Gb5* and its individual notes (**b**) *C4*, (**c**) *Gb5* y (**d**) *Eb6*

third degree (v.gr. *Eb6*) and its diminished fifth degree (v.gr. *Gb5*). If we compare the spectrum of *C4* (Fig. 2.12b) with the spectra of *Gb5* (Fig. 2.12c) and *Eb6* (Fig. 2.12d), it can be seen that there is no harmonic relation between these notes. So, these types of chord are easier to detect than, for example, the perfect major chord. In music, consonant intervals are more frequent that dissonant intervals. So, trained musicians find it more difficult to identify pitches of dissonant intervals than in consonant intervals [19]. However, if we compare the spectrum of the consonant chord represented in Fig. 2.11, with the spectrum of the dissonant chord represented in Fig. 2.12, it can be seen how in the consonant chord the overlap between harmonics makes it difficult to distinguish all the individual notes that compose this chord. On the other hand, in the dissonant chord the individual notes are a bit less difficult to separate.

It is noteworthy that the harmonics of the notes that compose a chord will be perfectly coincident if the inharmonicity of the piano strings didn't exist [23] and the piano were turned according to the just intonation scale. But the difference introduced by the inharmonicity of the piano strings and the equal tempered scale is so small that harmonic coincidence still happens.

Chapter 3
Description of the UMA Piano Chord Database

Abstract This chapter describes the content of the UMA piano chord database. First, the chords selected for the database and the naming convention of the played notes is presented. The naming convention used has musical meaning which is a new feature in this type of database. Finally, the content of the database organized by the polyphony number is shown.

3.1 Chord Selection

The chords are selected taking into account the most typical chords in Western music and also different difficulty levels in the detection due to the existing harmonic relations between the notes that compose the chord.

The selected chords are composed of 3–7 different notes. In the recording process, some of these notes are doubled to get the same chord but with different polyphony number. Table 3.1 shows the organization of the tables in which the different recorded chords are presented. This table includes references to the tables in Appendix A that contain, in addition to the information presented in the tables of this chapter, the MIDI difference between notes in the chords and musical examples. The different types of recorded chords with three, four, five, six and seven different notes are presented in Tables 3.2, 3.3, 3.4, 3.5 and 3.6, respectively.

In the tables indicated in Table 3.1, $X : (a,b,c,\ldots)$ denotes the different notes in the chords, where X, represent the root note of the chord, and $(a,b,c,\ldots)$ represent the basic interval with the root note. Note that, in the recordings, these intervals can be simple or composed by one or more octaves. The simple interval is represented by a number and an alteration (v.gr. 5 means a perfect fifth chord, $5b$ means a diminished fifth chord and 5# means augmented fifth chord (see Table 2.1))[15].

The most consonant chord that exist, and the most typical Western music chord, is the perfect major chord ($X : (1,3,5)$). The perfect major chord, in all the polyphony levels and with different frequency separation and order of the notes as well as with different doubled notes (v.gr. for polyphony number 4, we can double

A.M. Barbancho et al., *Database of Piano Chords: An Engineering View of Harmony*,   17
SpringerBriefs in Electrical and Computer Engineering,
DOI 10.1007/978-1-4614-7476-0_3, © The Author(s) 2013

Table 3.1 Organization of the tables in which the different recorded chords are presented

Number of notes	Tables
3	3.2, A.2
4	3.3, A.3
5	3.4, A.4, A.5
6	3.5, A.6, A.7
7	3.6, A.8, A.9

Table 3.2 Types of recorded chords with three different notes

Name	Intervals
Perfect Major	$X : (1,3,5)$
Perfect Minor	$X : (1,3b,5)$
Augmented	$X : (1,3,5\#)$
Augmented-Minor	$X : (1,3b,5\#)$
Diminished	$X : (1,3,5b)$
Diminished Minor	$X : (1,3b,5b)$

Table 3.3 Types of recorded chords with four different notes

Name	Intervals
Suspended 4	$X : (1,4,5,7)$
Suspended-Augmented 4	$X : (1,4\#,5,7)$
Suspended-Diminished 4	$X : (1,4b,5,7)$
Major-Major 7	$X : (1,3,5,7)$
Major-Minor 7	$X : (1,3,5,7b)$
Augmented-Major 7	$X : (1,3,5\#,7)$
Augmented-Minor 7	$X : (1,3,5\#,7b)$
Diminished-Major 7	$X : (1,3,5b,7)$
Diminished-Minor 7	$X : (1,3,5b,7b)$
Minor-Major 7	$X : (1,3b,5,7)$
Minor-Minor 7	$X : (1,3b,5,7b)$
Augmented-Minor-Major 7	$X : (1,3b,5\#,7)$
Augmented-Minor-Minor 7	$X : (1,3b,5\#,7b)$
Diminished-Minor-Major 7	$X : (1,3b,5b,7)$
Diminished-Minor-Minor 7	$X : (1,3b,5b,7b)$

the root note, the third degree note and the fifth degree note) are included in the designed UMA-Piano database. These chords are considered difficult to detect.

Recall that the relative positions between the notes and the piano octave in which a chord is recorded are important (see Sect. 2.3).

Also note the two different groups of chords in Western music [13]. In the dissonant chords, the piano octave in which chords are recorded as well as doubled notes, make chords easier or more difficult to detect.

Therefore, in the UMA-Piano database, the most typical chords in Western music, evolving through different difficulty levels of detection because of the harmonic relation between the notes that compose the chord and the different frequency separation and order of the notes, have been included.

Table 3.4 Types of recorded chords with five different notes

Name	Intervals
Major-Major 7-Major 9	$X : (1,3,5,7,9)$
Major-Major 7-Minor 9	$X : (1,3,5,7,9b)$
Major-Minor 7-Major 9	$X : (1,3,5,7b,9)$
Major-Minor 7-Minor 9	$X : (1,3,5,7b,9b)$
Augmented-Major 7-Major 9	$X : (1,3,5\#,7,9)$
Augmented-Major 7-Minor 9	$X : (1,3,5\#,7,9b)$
Augmented-Minor 7-Major 9	$X : (1,3,5\#,7b,9)$
Augmented-Minor 7-Minor 9	$X : (1,3,5\#,7b,9b)$
Diminished-Major 7-Major 9	$X : (1,3,5b,7,9)$
Diminished-Major 7-Minor 9	$X : (1,3,5b,7,9b)$
Diminished-Minor 7-Major 9	$X : (1,3,5b,7b,9)$
Diminished-Minor 7-Minor 9	$X : (1,3,5b,7b,9b)$
Minor-Major 7-Major 9	$X : (1,3b,5,7,9)$
Minor-Major 7-Minor 9	$X : (1,3b,5,7,9b)$
Minor-Minor 7-Major 9	$X : (1,3b,5,7b,9)$
Minor-Minor 7-Minor 9	$X : (1,3b,5,7b,9b)$
Augmented-Minor-Major 7-Major 9	$X : (1,3b,5\#,7,9)$
Augmented-Minor-Major 7-Minor 9	$X : (1,3b,5\#,7,9b)$
Augmented-Minor-Minor 7-Major 9	$X : (1,3b,5\#,7b,9)$
Augmented-Minor-Minor 7-Minor 9	$X : (1,3b,5\#,7b,9b)$
Diminished-Minor-Major 7-Major 9	$X : (1,3b,5b,7,9)$
Diminished-Minor-Major 7-Minor 9	$X : (1,3b,5b,7,9b)$
Diminished-Minor-Minor 7-Major 9	$X : (1,3b,5b,7b,9)$
Diminished-Minor-Minor 7-Minor 9	$X : (1,3b,5b,7b,9b)$

3.2 Naming Convention of Played Notes

The names of the notes played are designed as the name of the note with or without alteration and the corresponding musical index (v.gr. $Gb4$, $G\#4$, $G4$). As it has been said before, we consider that the note $A4$ correspond to a pitch of 440 Hz [12].

Due to the existence of enharmonic sound in the equal temperament scale (Fig. 2.5), which means that notes sharing the same in pitch bear different names (v.gr $C\#$ and Db, $D\#$ and Eb, etc.), we need to decide which name to use for each note.

In music, as described in Sect. 2.2, the tonality is the harmonic center of gravity of a musical excerpt. Each tonality or key has a key signature or number of alterations. These alterations can be sharps (#) or flats (b).

From the musicians point of view, it is important to code the name of the notes with sharps or flats depending of the key of the musical piece. For example, in a musical piece in C minor, a perfect minor chord is represented by $C4Eb4G4$. In this case, the perfect minor chord can also be represented by $C4D\#4G4$ ($Eb4$ is enharmonic with $D\#4$), but it is not its normal representation and, so, it can be a source of error.

Therefore, in the UMA-Piano chord database, we name the notes (with sharps or flats) in the more common way for each of the considered keys (v.gr. In C minor Bb,

Table 3.5 Types of recorded chords with six different notes

Name	Intervals
Major-Major 7-Major 9-Perfect 11	$X : (1,3,5,7,9,11)$
Major-Major 7-Minor 9-Perfect 11	$X : (1,3,5,7,9b,11)$
Major-Minor 7-Major 9-Perfect 11	$X : (1,3,5,7b,9,11)$
Major-Minor 7-Minor 9-Perfect 11	$X : (1,3,5,7b,9b,11)$
Augmented-Major 7-Major 9-Perfect 11	$X : (1,3,5\#,7,9,11)$
Augmented-Major 7-Minor 9-Perfect 11	$X : (1,3,5\#,7,9b,11)$
Augmented-Minor 7-Major 9-Perfect 11	$X : (1,3,5\#,7b,9,11)$
Augmented-Minor 7-Minor 9-Perfect 11	$X : (1,3,5\#,7b,9b,11)$
Diminished-Major 7-Major 9-Perfect 11	$X : (1,3,5b,7,9,11)$
Diminished-Major 7-Minor 9-Perfect 11	$X : (1,3,5b,7,9b,11)$
Diminished-Minor 7-Major 9-Perfect 11	$X : (1,3,5b,7b,9,11)$
Diminished-Minor 7-Minor 9-Perfect 11	$X : (1,3,5b,7b,9b,11)$
Minor-Major 7-Major 9-Perfect 11	$X : (1,3b,5,7,9,11)$
Minor-Major 7-Minor 9-Perfect 11	$X : (1,3b,5,7,9b,11)$
Minor-Minor 7-Major 9-Perfect 11	$X : (1,3b,5,7b,9,11)$
Minor-Minor 7-Minor 9-Perfect 11	$X : (1,3b,5,7b,9b,11)$
Augmented-Minor-Major 7-Major 9-Perfect 11	$X : (1,3b,5\#,7,9,11)$
Augmented-Minor-Major 7-Minor 9-Perfect 11	$X : (1,3b,5\#,7,9b,11)$
Augmented-Minor-Minor 7-Major 9-Perfect 11	$X : (1,3b,5\#,7b,9,11)$
Augmented-Minor-Minor 7-Minor 9-Perfect 11	$X : (1,3b,5\#,7b,9b,11)$
Diminished-Minor-Major 7-Major 9-Perfect 11	$X : (1,3b,5b,7,9,11)$
Diminished-Minor-Major 7-Minor 9-Perfect 11	$X : (1,3b,5b,7,9b,11)$
Diminished-Minor-Minor 7-Major 9-Perfect 11	$X : (1,3b,5b,7b,9,11)$
Diminished-Minor-Minor 7-Minor 9-Perfect 11	$X : (1,3b,5b,7b,9b,11)$

Eb, *Ab* and in A Major *F#*, *C#*, *G#*) [13]. This characteristic is very important from the musician point of view and should help to relate the musical point of view to the signal processing one. An example of the name of the notes for the different scale degrees for the *C* key is presented in Table 3.7. The names of the notes in this table are the most usual ones when a piece is in *C* key (major or minor).

3.3   Content of the UMA-Piano Chord Database

The UMA-Piano chord database presented contains chords with polyphony number ranging from 2 to 10 and individual recordings of each of the 88 piano keys. All the recordings are provided in three dynamics levels (*Forte, Mezzo, Piano*) and three different playing styles (*Normal, Staccato, Pedal*). These chord recordings can be used without requiring any preprocessing like dividing each file in different chords, unlike the RWC database [9].

In Fig. 3.1, the naming convention of sound files is presented. The naming convention allows to identify each chord uniquely. The names of the notes played are designed as the name of the note with or without alteration and its corresponding musical index (v.gr. *Gb4*, *G#4*, *G4*).

Table 3.6 Types of recorded chords with seven different notes

Name	Intervals
Major-Major 7-Major 9-Perfect 11-Major 13	$X:(1,3,5,7,9,11,13)$
Major-Major 7-Minor 9-Perfect 11-Major 13	$X:(1,3,5,7,9b,11,13)$
Major-Minor 7-Major 9-Perfect 11-Major 13	$X:(1,3,5,7b,9,11,13)$
Major-Minor 7-Minor 9-Perfect 11-Major 13	$X:(1,3,5,7b,9b,11,13)$
Augmented-Major 7-Major 9-Perfect 11-Major 13	$X:(1,3,5\#,7,9,11,13)$
Augmented-Major 7-Minor 9-Perfect 11-Major 13	$X:(1,3,5\#,7,9b,11,13)$
Augmented-Minor 7-Major 9-Perfect 11-Major 13	$X:(1,3,5\#,7b,9,11,13)$
Augmented-Minor 7-Minor 9-Perfect 11-Major 13	$X:(1,3,5\#,7b,9b,11,13)$
Diminished-Major 7-Major 9-Perfect 11-Major 13	$X:(1,3,5b,7,9,11,13)$
Diminished-Major 7-Minor 9-Perfect 11-Major 13	$X:(1,3,5b,7,9b,11,13)$
Diminished-Minor 7-Major 9-Perfect 11-Major 13	$X:(1,3,5b,7b,9,11,13)$
Diminished-Minor 7-Minor 9-Perfect 11-Major 13	$X:(1,3,5b,7b,9b,11,13)$
Minor-Major 7-Major 9-Perfect 11-Major 13	$X:(1,3b,5,7,9,11,13)$
Minor-Major 7-Minor 9-Perfect 11-Major 13	$X:(1,3b,5,7,9b,11,13)$
Minor-Minor 7-Major 9-Perfect 11-Major 13	$X:(1,3b,5,7b,9,11,13)$
Minor-Minor 7-Minor 9-Perfect 11-Major 13	$X:(1,3b,5,7b,9b,11,13)$
Augmented-Minor-Major 7-Major 9-Perfect 11-Major 13	$X:(1,3b,5\#,7,9,11,13)$
Augmented-Minor-Major 7-Minor 9-Perfect 11-Major 13	$X:(1,3b,5\#,7,9b,11,13)$
Augmented-Minor-Minor 7-Major 9-Perfect 11-Major 13	$X:(1,3b,5\#,7b,9,11,13)$
Augmented Minor-Minor 7-Minor 9-Perfect 11-Major 13	$X:(1,3b,5\#,7b,9b,11,13)$
Diminished-Minor-Major 7-Major 9-Perfect 11-Major 13	$X:(1,3b,5b,7,9,11,13)$
Diminished-Minor-Major 7-Minor 9-Perfect 11-Major 13	$X:(1,3b,5b,7,9b,11,13)$
Diminished-Minor-Minor 7-Major 9-Perfect 11-Major 13	$X:(1,3b,5b,7b,9,11,13)$
Diminished-Minor-Minor 7-Minor 9-Perfect 11-Major 13	$X:(1,3b,5b,7b,9b,11,13)$

Table 3.7 Names of the notes for the different scale degrees in C key

Degree	I	II	III	IV	V	VI	VII
Degree name	Tonic	Supertonic	Mediant	Subdominant	Dominant	Superdominant	Subtonic
Note name	C	D	E	F	G	A	B

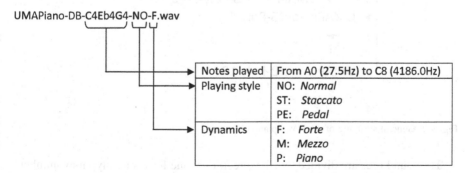

UMAPiano-DB-C4Eb4G4-NO-F.wav

Notes played	From A0 (27.5Hz) to C8 (4186.0Hz)
Playing style	NO: *Normal* ST: *Staccato* PE: *Pedal*
Dynamics	F: *Forte* M: *Mezzo* P: *Piano*

Fig. 3.1 Naming convention of sound files

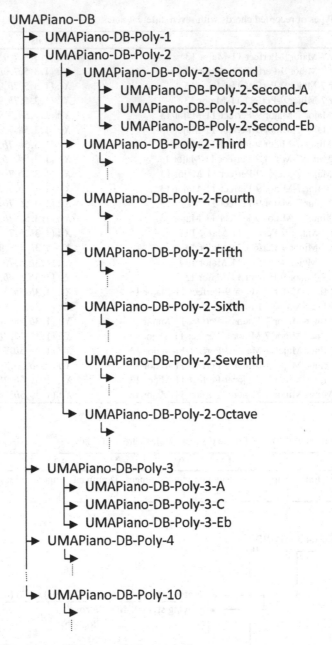

Fig. 3.2 General structure of the UMA piano database

The sound files are divided into 10 directories, one for each polyphony number. In Fig. 3.2, the general structure of the database is presented. The name of each directory is 'UMAPiano-DB-Poly-Y', where Y goes from 1 to 10.

Table 3.8 Directories in '_UMAPiano-DB-Poly-2_'

UMAPiano-DB-Poly-2-Second
UMAPiano-DB-Poly-2-Third
UMAPiano-DB-Poly-2-Fourth
UMAPiano-DB-Poly-2-Fifth
UMAPiano-DB-Poly-2-Sixth
UMAPiano-DB-Poly-2-Seventh
UMAPiano-DB-Poly-2-Octave

Table 3.9 Directories in '_UMAPiano-DB-Poly-2-X_', where '_X_' can be: Second, third, fourth, fifth, sixth, seventh or octave

UMAPiano-DB-Poly-2-X-A
UMAPiano-DB-Poly-2-X-C
UMAPiano-DB-Poly-2-X-Eb

Table 3.10 Directories in '_UMAPiano-DB-Poly-3_'

UMAPiano-DB-Poly-3-A
UMAPiano-DB-Poly-3-C
UMAPiano-DB-Poly-3-Eb

Each of these directories is also divided into several directories. For polyphony 2 recordings, due to the large number of recordings, seven directories have been created, that are related to the interval between the two notes that compose each chord. In Table 3.8, the names of the directories for polyphony 2 are presented. Furthermore, each of the directories containing polyphony 2 recordings, is divided into three directories, recording to the bass note. In Table 3.9 the names of these directories are presented.

The recordings with polyphony 3–10 are organized into directories related to the fundamental tone of the key of the recorded chord. In Table 3.10 an example of this naming convention for the name of the directories for polyphony 3 is presented.

For all the polyphony numbers, all the chords have been recorded considering the keys: _C_ Major or minor, _A_ Major or minor and _Eb_ Major or minor.

In the following sections, we give a detailed description of the content of the six main directories of the database.

3.3.1 Single Note Recordings

In the directory '_UMAPiano-DB-Poly-1_' the individual recordings of each of the 88 piano keys are contained. The total number of files is 88×3 _dynamics_ $\times 3$ _styles_ $= 792$. The naming of the notes consider all the altered notes as sharps.

Table 3.11 Kinds of intervals considered for polyphony 2 recordings

Intervals	Kinds			
Second	Minor	Major		
Third	Minor	Major		
Fourth	Perfect			
Fifth	Diminished	Perfect	Augmented	
Sixth	Major			
Seventh	Minor	Major		
Octave	Perfect			

3.3.2 Polyphony 2 Recordings

In the directory 'UMAPiano-DB-Poly-2' the recordings of chords with polyphony 2 can be found. The total number of files is 7,128. Due to the large number of recordings, these are organized into seven directories, that correspond to the interval between the two notes that compose each chord. Moreover each directory, is divided into three directories that correspond to different bass notes.

The bass notes of the chord of polyphony 2, are the tonic of the recorded keys (C, A, Eb).

For all the recorded intervals, we have considered the simple interval (v.gr. $C1E1$) as well as the compound intervals from 1 to 3 octaves (v.gr. $C1E2, C1E3, C1E4$).

We have also considered all the second intervals in all the possible octaves of the piano (v.gr. $C1E1, C2E2, C3E3, \ldots, C7E7, \ldots C1E4, C2E5, C3E6, C4E7$).

Like the single note recordings, all these intervals are recorded with three dynamics levels and three different playing styles.

In Table 3.11, the kinds of intervals considered for each recording are presented. Note that these intervals include all the possible intervals avoiding repetitions (v.gr. Augmented seconds (3 semitones) = Minor Third (3 semitones)).

All these recording are useful to check the performance of transcription systems with all the possible intervals of seconds, as well as to define spectral patterns for transcription purposes [7, 8].

3.3.3 Polyphony 3 Recordings

In the directory 'UMAPiano-DB-Poly-3' the recordings of chords with polyphony 3 are contained. The total number of files is 19,503.

The recorded polyphony 3 chords can be divided into six groups. An example of chords of each group is shown in Fig. 3.3. The chords presented in this figure are in C Major key. For Groups 1–3, Fig. 3.3a–c represent the Perfect Major chord, as example of the chords in these groups (Perfect minor, Augmented, etc.). For Group 4, Fig. 3.3d represents the suspended four chord without fifth degree, as example of the chords in these groups For Groups 5 and 6, Fig. 3.3e, f represent the perfect major chord.

Fig. 3.3 Basic chords with polyphony 3. (**a**) Group 1. (**b**) Group 2. (**c**) Group 3. (**d**) Group 4. (**e**) Group 5. (**f**) Group 6

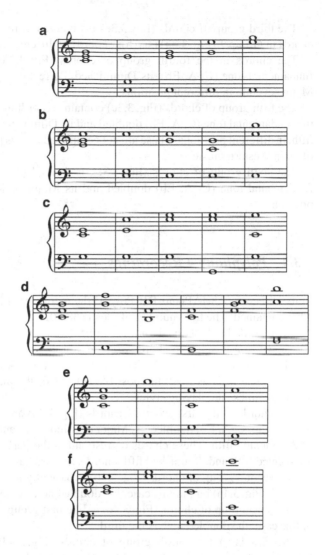

All these chords are recorded in *C* Major and minor, *A* Major and minor, *Eb* Major and minor keys as well as in all the possible octaves of the piano keyboard and in the three dynamics and playing styles.

The first group of chords (Fig. 3.3a), is composed by the fundamental tone (*C*, *A*, *Eb*), its Major or Minor third and its Diminished, Perfect or Augmented fifth. These chords correspond to the perfect Major, perfect Minor, augmented fifth and diminished fifth chords, which are basic in Western music. The main difference among the chords in Group 1 is the separation between notes.

The second group of chords (Fig. 3.3b) corresponds to the first inversion of the chords in Group 1 (Fig. 3.3a) (sixth chord), in different positions.

The third group of chords (Fig. 3.3c) is obtained as the second inversion of the chords in Group 1 (Fig. 3.3a) (six-four chord) in different positions.

The chords in the fourth group of chords (Fig. 3.3d) are composed by the fundamental tone (C, A, Eb), its Diminished, Perfect or Augmented fourth and its Major seventh, in different positions.

The fifth group of chords (Fig. 3.3e) contains recordings of chords composed by the fundamental tone (C, A, Eb) doubled and its Diminished, Perfect or Augmented fifth, in different positions. These are not completed chords [13], these chords appear often in Western music.

The sixth group of chords (Fig. 3.3f), is formed by chords composed by the fundamental tone (C, A, Eb) doubled and its Major or Minor third, in different positions.

3.3.4   Polyphony 4 Recordings

In the directory 'UMAPiano-DB-Poly-4' the recordings of chords with polyphony 4 are contained. The total number of files is 38,502.

The recorded polyphony 4 chords can be divided into five groups. These groups are shown in Fig. 3.4. The chords presented in this figure are in C Major key. All the chords are recorded in C Major, C minor, A Major, A minor, Eb Major and Eb minor keys as well as in all the possible octaves of the piano keyboard and in the three dynamics and playing styles.

The chords in the first group of chords (Fig. 3.4a) are composed by the fundamental tone (C, A, Eb) doubled, its Major or Minor third and its Diminished, Perfect or Augmented fifth. These chords correspond to the perfect Major, perfect Minor, augmented fifth and diminished fifth chords, that are basic in Western music.

The second group of chords (Fig. 3.4b) is basically the same group of chords as Group 1 (Fig. 3.4a) but, in this case, the doubled note is the fifth.

The third group of chords (Fig. 3.4c) is like first group of chords (Fig. 3.4a) but, in this case, the doubled note is the third.

The chords in the fourth group of chords (Fig. 3.4d) are composed by the fundamental tone (C, A, Eb), its Major or Minor third, its Diminished, Perfect or Augmented fifth and its Major or Minor seventh. These chords correspond to seventh chords, that constitute another group of typical chords in Western music.

The fifth group of chords (Fig. 3.4e) contains similar chords as group 4 (Fig. 3.4d) but, in this case, the seventh is changed by a Major or Minor ninth. These chords correspond to ninth chords.

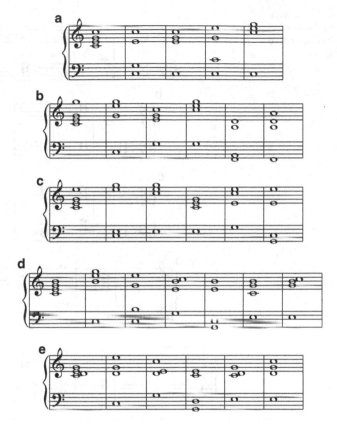

Fig. 3.4 Basic chords with polyphony 4. (**a**) Group 1. (**b**) Group 2. (**c**) Group 3. (**d**) Group 4. (**e**) Group 5

3.3.5 Polyphony 5 Recordings

In the directory 'UMAPiano-DB-Poly-5' the recordings of chords with polyphony 5 are contained. The total number of files is 33,948.

The recorded polyphony 5 chords can be divided into the five groups shown in Fig. 3.5. The chords presented in this figure are in C Major key. All these chords are recorded in C Major, C minor, A Major, A minor, Eb Major and Eb minor keys as well as in all the possible octaves of the piano keyboard and in the three dynamics and playing styles.

The first group of chords (Fig. 3.5a) contains chords composed by the fundamental tone (C, A, Eb) doubled, its Major or Minor third and its Diminished, Perfect or Augmented fifth doubled. These chords correspond to the perfect Major, perfect Minor, augmented fifth and diminished fifth chords, that are basic in Western music with double of the fundamental tone and its fifth.

Fig. 3.5 Basic chords with
polyphony 5. (**a**) Group 1.
(**b**) Group 2. (**c**) Group 3. (**d**)
Group 4. (**e**) Group 5

The second group of chords (Fig. 3.5b) is similar to Group 1 (Fig. 3.5a) but, in this case, the doubled notes are the fifth and the third of the fundamental tone.

The third group of chords (Fig. 3.5c) is also similar to Group 1 (Fig. 3.5a) but, in this case, the doubled notes are the fundamental tone and its third.

The chords in the fourth group of chords (Fig. 3.5d) are composed by the fundamental tone (C, A, Eb) doubled, its Major or Minor third, its Diminished, Perfect or Augmented fifth and its Major or Minor seventh. These chords correspond to seventh chords. This is a group of chords commonly used in Western music.

The fifth group of chords (Fig. 3.5e) is defined similarly to Group 4 (Fig. 3.5d) but, in this case, the seventh is changed by a Major or Minor ninth and the fundamental is doubled. These chords correspond to ninth chords.

3.3.6   Polyphony 6 Recordings

In the directory 'UMAPiano-DB-Poly-6' the recordings of chords with polyphony 6 are contained. The total number of files is 51,777.

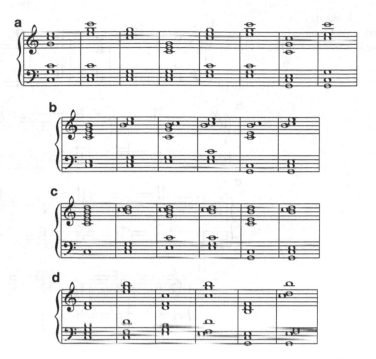

Fig. 3.6 Basic chords with polyphony 6. (**a**) Group 1. (**b**) Group 2. (**c**) Group 3. (**d**) Group 4

The recorded polyphony 6 chords can be divided into the four groups that are shown in Fig. 3.6. The chords presented in this figure are in C Major key. All the chords are recorded in C Major, C minor, A Major, A minor, Eb Major and Eb minor keys as well as in all the possible octaves of the piano keyboard and in the three dynamics and playing styles.

The chords in the first group of chords (Fig. 3.6a) are composed by the fundamental tone (C, A, Eb) doubled, its Major or Minor third doubled and its Diminished, Perfect or Augmented fifth doubled. These chords correspond to the perfect Major, perfect Minor, augmented fifth and diminished fifth chords.

The second group of chords (Fig. 3.6b) contains chords composed by the fundamental tone (C, A, Eb) doubled, its Major or Minor third, its Diminished, Perfect or Augmented fifth, its Major or Minor seventh and its Major or Minor ninth. These chords correspond to ninth chords with seventh, that are another group of common chords in Western music.

The third group of chords (Fig. 3.6c) is formed by chords composed by the fundamental tone (C, A, Eb), its Major or Minor third, its Diminished, Perfect or Augmented fifth, its Major or Minor seventh, its Major ninth and its perfect eleventh. These chords correspond to 11th chords, which is a group of commonly used chords in Western and jazz music.

Fig. 3.7 Basic chords with
polyphony 7. (**a**) Group 1.
(**b**) Group 2. (**c**) Group 3. (**d**)
Group 4

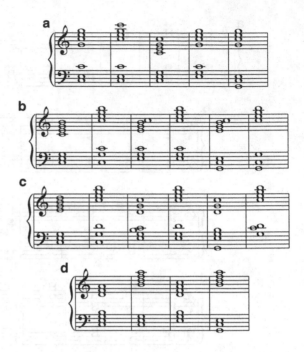

The content of the fourth group of chords, presented in Fig. 3.6d, is slightly
different from the set of chords in the other groups. In this case, the recorded chords
solely contain the intervals presented, with no options of different intervals (v.gr.
Major or Minor third, etc.). These chords are the mixture of two perfect major or
perfect minor chord with different fundamental tone. These chords are useful to test
the behaviour of transcription systems when there is no harmonical relation between
the notes in a chord.

3.3.7 Polyphony 7 Recordings

In the directory 'UMAPiano-DB-Poly-7' the recordings of chords with polyphony
7 are contained. The total number of files is 45,954.

The recorded polyphony 7 chords can be divided into the four groups shown in
Fig. 3.7. The chords presented in this figure are in C Major key. All these chords are
recorded in C Major, C minor, A Major, A minor, Eb Major and Eb minor keys, as
well as in all the possible octaves of the piano keyboard and in the three dynamics
and playing styles.

The first group of chords (Fig. 3.7a) contains chords composed by the fundamental tone (C, A, Eb) doubled, its Major or Minor third doubled and its Diminished,
Perfect or Augmented fifth doubled. These chords correspond to the perfect Major,
perfect Minor, augmented fifth and diminished fifth chords.

Fig. 3.8 Basic chords with
polyphony 8. (**a**) Group 1.
(**b**) Group 2. (**c**) Group 3

The chords in the second group (Fig. 3.7b) are composed by the fundamental tone
(C, A, Eb) doubled, its Major or Minor third, its Diminished, Perfect or Augmented
fifth and its Major or Minor seventh. These chords correspond to seventh chords,
commonly used in Western music.

In the third group of chords (Fig. 3.7c), these are composed by the fundamental
tone (C, A, Eb), its Major or Minor third, its Diminished, Perfect or Augmented fifth,
its Major or Minor seventh and its Major or Minor ninth. These chords correspond
to ninth chords with seven, which are common in Western and jazz music.

The chords in the fourth group (Fig. 3.7c) are composed by the fundamental tone
(C, A, Eb), its Major or Minor third, its Diminished, Perfect or Augmented fifth,
its Major or Minor seventh, its Major or Minor ninth, its perfect eleventh and its
Major thirteenth. These chords correspond to 13th chords, which are common in
jazz music.

3.3.8   Polyphony 8 Recordings

In the directory 'UMAPiano-DB-Poly-8' the recordings of chords with polyphony
8 are contained. The total number of files is 31,536.

The recorded polyphony 8 chords can be divided into three groups. These groups
are shown in Fig. 3.8. The chords presented in this figure are in C Major key. All
these chords are recorded in C Major, C minor, A Major, A minor, Eb Major and Eb
minor keys as well as in all the possible octaves of the piano keyboard and in the
three dynamics and playing styles.

The chords in the first group of (Fig. 3.8a) are composed by the fundamental tone
(C, A, Eb) doubled, its Major or Minor third doubled and its Diminished, Perfect
or Augmented fifth doubled. These chords correspond to the perfect Major, perfect
Minor, augmented fifth and diminished fifth chords.

Fig. 3.9 Basic chords with
polyphony 9. (**a**) Group 1. (**b**)
Group 2

The second group of chords (Fig. 3.8b) contains chords composed by the fundamental tone (C, A, Eb) doubled, its Major or Minor third, its Diminished, Perfect or Augmented fifth and its Major or Minor seventh. These chords correspond to seventh chords, which is another group of typical chords in Western music.

The third group of chords (Fig. 3.8c) contains recordings composed by the fundamental tone (C, A, Eb), its Major or Minor third, its Diminished, Perfect or Augmented fifth, its Major or Minor seventh, its Major or Minor ninth, its perfect eleventh and its Major thirteenth. These chords correspond to 13th chords which is a group of commonly used chords in jazz music.

3.3.9   Polyphony 9 Recordings

In the directory 'UMAPiano-DB-Poly-9' the recordings of chords with polyphony 9 are contained. The total number of files is 22,140.

The recorded polyphony 9 chords can be divided into the two groups shown in Fig. 3.9. The chords presented in this figure are in C Major key. All these chords are recorded in C Major, C minor, A Major, A minor, Eb Major and Eb minor keys as well as in all the possible octaves of the piano keyboard and in the three dynamics and playing styles.

The first group contains chords (Fig. 3.9a) composed by the fundamental tone (C, A, Eb) doubled, its Major or Minor third, its Diminished, Perfect or Augmented fifth and its Major or Minor seventh. These chords correspond to seventh chords which constitute group of common chords in Western music.

The chords in the second group (Fig. 3.8b) are composed by the fundamental tone (C, A, Eb), its Major or Minor third, its Diminished, Perfect or Augmented fifth, its Major or Minor seventh, its Major or Minor ninth, its perfect eleventh and its Major thirteenth. These chords correspond to 13th chords which is a typical group of chords in jazz music.

Fig. 3.10 Basic chords with
polyphony 10. (**a**) Group 1.
(**b**) Group 2

3.3.10 Polyphony 10 Recordings

In the directory 'UMAPiano-DB-Poly-10' the recordings of chords with polyphony
9 are contained. The total number of files is 23,760.

The recorded polyphony 9 chords can be divided into the two groups shown in
Fig. 3.10. The chords presented in this figure are in C Major key. All these chords
are recorded in C Major, C minor, A Major, A minor, Eb Major and Eb minor keys
as well as in all the possible octaves of the piano keyboard and in the three dynamics
and playing styles.

The chords in the first group (Fig. 3.10a) are composed by the fundamental tone
(C, A, Eb) doubled, its Major or Minor third, its Diminished, Perfect or Augmented
fifth and its Major or Minor seventh. These chords correspond to seventh chords.
These chords are commonly used in Western music.

In the second group, the chords (Fig. 3.10b) are composed by the fundamental
tone (C, A, Eb), its Major or Minor third, its Diminished, Perfect or Augmented
fifth, its Major or Minor seventh, its Major or Minor ninth, its perfect eleventh and its
Major thirteenth. These chords correspond to 13th chords, which are typical chords
in jazz music.

3.4 Descriptions of the Recordings

The sounds of the piano were recorded at a frequency rate of 44.1 kHz and quantized
with 16 bits (standard CD quality). The recordings are stored in 275,040 monaural
sound files with a total size of about 54 GB and a total play back time of about 170 h.
The processes of recordings were done on a professional digital audio production
system. The piano used for the chord recordings is a Kawai CA91.

In order to take advantage of the full dynamic range of the 16 bit A/D converter
used, different gains have been applied to the chords belonging to different dynamics
(*Forte, Mezzo, Piano*). These gains have been chosen in such way that louder chords
do not clip, but they still cover almost the full dynamic range. The louder dynamic,
Forte, has not been amplified, it is used as reference and so it is considered to be

recorded with a gain equal to 0 dB. The other dynamics have been amplified 8 dB in the case of *Mezzo* and 24 dB in the case of *Piano*. This feature should be considered before a direct comparison of amplitudes between the various dynamics.

Each file, contains a single chord or a single note, as we presented in the beginning of Sect. 3.3. The naming of the recorded chord is the more common for each of the recorded keys (*C* Major, *C* minor, *A* Major, *A* minor, *Eb* Major and *Eb* minor) (see Sect. 3.2).

This database will be available to the researching community on Springer's Extra Materials website.

Chapter 4
Summary and Discussion

Abstract This chapter presents the summary and the discussion of this book and database. In this book the UMA-Piano Chord Database is presented as well as an engineering point of view of harmony. This book ends with a discussion of the future improvements of the database presented.

4.1 Summary

In this book the UMA-Piano chord Database is made available and described in detail. This data base is available to researchers around the world at www. SpringerLink.com.

The designed database contains chords with polyphony number ranging from 2 to 10 and individual recordings of each of the 88 piano keys.

All the recordings of both the notes and the chords are done in three dynamics (*Forte, Mezzo, Piano*) and three different playing styles (*Normal, Staccato, Pedal*). Each dynamic (*Forte, Mezzo, Piano*) has been separately recorded with a different gain in order to take advantage of the full recording dynamic range. The applied amplification is 0 dB for *Forte*, 8 dB for *Mezzo* and 24 dB for *Piano*.

The recorded chords correspond to the types of chords actually used in Western music [13].

The total number of recorded files is 275,040, so this is a large database that contains a very significant number of chord of all the types with a size over 54 GB and a total play back time of about 170 h.

This book also contains an engineering point of view of harmony that allows to understand the content of the database as well as the foundations of Western music for those lay in musical concepts.

This book also attempts to fill in the gap between common musical notation and engineering and signal processing concepts including descriptions and notation.

A.M. Barbancho et al., *Database of Piano Chords: An Engineering View of Harmony,*   35
SpringerBriefs in Electrical and Computer Engineering,
DOI 10.1007/978-1-4614-7476-0_4, © The Author(s) 2013

An appendix (Appendix A) with lots of examples of chords including musical notation and the description of the chords in musical intervals, MIDI numbers and scores is given.

The descriptions included in this book to build the database can be used to create complete databases of other instruments with main focus on the musical meaning of the database in Western music. Please mention that the UMA Piano chord Database has been used when reporting research results (papers, publications, software or any other outcome) obtained using the database described. Also include a clear citation to this book.

4.2 Discussion

In the future, this UMA-Piano database can be improved in the following ways:

- Increasing the number of pianos used for the chord recordings: it would be desirable to have recordings of other different pianos to avoid the dependence on a specific piano.
- Augmenting the considered set of keys to get the 12 Major keys and the 12 Minor keys.

This would complete the set of possible chords with musical meaning, greatly increasing the size of the database.

In spite of these improvements, this database is a complete database of notes and chords with musical meaning focused on Western music.

Appendix A
Types of Recorded Chords

In this appendix, detailed lists of the types of recorded chords are presented. These lists include:

- The conventional name of the chord [13, 15].
- The intervals between the notes that compose the chord. The intervals are denoted as $X : (a, b, c, \ldots)$ where X represents the root note of the chord, and $(a, b, c, \ldots)$ represent the basic interval with the root note. Note that, in the recordings, these intervals can be simple or composed by one or more octaves. These intervals are represented by a number and an alteration (v.gr. 5 means a perfect fifth chord, $5b$ means a diminished fifth chord and $5\#$ means augmented fifth chord (see Table 2.1)) [15].
- The intervals between the notes that compose the chord expressed in MIDI numbers.
- An example of the chord for the C key.

The selected chords contain from 3 to 7 different tones. Note that some chords contain doubled notes, then up to 10 different sounds can be contained in the recorded chords (see Sect. 3.3)

In the recording process, some of these notes are doubled to give rise to the same chord but with different polyphony number. Table A.1 shows the organization of the tables in which the different recorded chords are presented. This table also refers the tables in Sect. 3.1.

A.M. Barbancho et al., *Database of Piano Chords: An Engineering View of Harmony*, 37
SpringerBriefs in Electrical and Computer Engineering,
DOI 10.1007/978-1-4614-7476-0, © The Author(s) 2013

Table A.1 Organization of
the tables in which the
different recorded chords are
presented

Number of notes	Tables
3	3.2, A.2
4	3.3, A.3
5	3.4, A.4, A.5
6	3.5, A.6, A.7
7	3.6, A.8, A.9

Table A.2 Types of recorded chords with three different notes

Name	Intervals	MIDI Intervals	Example
Perfect Major	$X : (1,3,5)$	$MIDI : (0,4,7)$	
Perfect Minor	$X : (1,3b,5)$	$MIDI : (0,3,7)$	
Augmented	$X : (1,3,5\#)$	$MIDI : (0,4,8)$	
Augmented - Minor	$X : (1,3b,5\#)$	$MIDI : (0,3,8)$	
Diminished	$X : (1,3,5b)$	$MIDI : (0,4,6)$	
Diminished - Minor	$X : (1,3b,5b)$	$MIDI : (0,3,6)$	

Table A.3 Types of recorded chords with four different notes

Name	Intervals	MIDI Intervals	Example
Suspended 4	$X : (1,4,5,7)$	$MIDI : (0,5,7,11)$	
Suspended - Augmented 4	$X : (1,4\#,5,7)$	$MIDI : (0,6,7,11)$	
Suspended - Diminished 4	$X : (1,4b,5,7)$	$MIDI : (0,4,7,11)$	
Major - Major 7	$X : (1,3,5,7)$	$MIDI : (0,4,7,11)$	
Major - Minor 7	$X : (1,3,5,7b)$	$MIDI : (0,4,7,10)$	
Augmented - Major 7	$X : (1,3,5\#,7)$	$MIDI : (0,4,8,11)$	
Augmented - Minor 7	$X : (1,3,5\#,7b)$	$MIDI : (0,4,8,10)$	
Diminished - Major 7	$X : (1,3,5b,7)$	$MIDI : (0,4,6,11)$	
Diminished - Minor 7	$X : (1,3,5b,7b)$	$MIDI : (0,4,6,10)$	
Minor - Major 7	$X : (1,3b,5,7)$	$MIDI : (0,3,7,11)$	
Minor - Minor 7	$X : (1,3b,5,7b)$	$MIDI : (0,3,7,10)$	
Augmented - Minor - Major 7	$X : (1,3b,5\#,7)$	$MIDI : (0,3,8,11)$	
Augmented - Minor - Minor 7	$X : (1,3b,5\#,7b)$	$MIDI : (0,3,8,10)$	
Diminished - Minor - Major 7	$X : (1,3b,5b,7)$	$MIDI : (0,3,6,11)$	
Diminished - Minor - Minor 7	$X : (1,3b,5b,7b)$	$MIDI : (0,3,6,10)$	

Table A.4 Types of recorded chords with five different notes (I)

Name	Intervals	MIDI Intervals	Example
Major - Major 7 - Major 9	$X : (1,3,5,7,9)$	$MIDI : (0,4,7,11,14)$	
Major - Major 7 - Minor 9	$X : (1,3,5,7,9b)$	$MIDI : (0,4,7,11,13)$	
Major - Minor 7 - Major 9	$X : (1,3,5,7b,9)$	$MIDI : (0,4,7,10,14)$	
Major - Minor 7 - Minor 9	$X : (1,3,5,7b,9b)$	$MIDI : (0,4,7,10,13)$	
Augmented - Major 7 - Major 9	$X : (1,3,5\#,7,9)$	$MIDI : (0,4,8,11,14)$	
Augmented - Major 7 - Minor 9	$X : (1,3,5\#,7,9b)$	$MIDI : (0,4,8,11,13)$	
Augmented - Minor 7 - Major 9	$X : (1,3,5\#,7b,9)$	$MIDI : (0,4,8,10,14)$	
Augmented - Minor 7 - Minor 9	$X : (1,3,5\#,7b,9b)$	$MIDI : (0,4,8,10,13)$	
Diminished - Major 7 - Major 9	$X : (1,3,5b,7,9)$	$MIDI : (0,4,6,11,14)$	
Diminished - Major 7 - Minor 9	$X : (1,3,5b,7,9b)$	$MIDI : (0,4,6,11,13)$	
Diminished - Minor 7 - Major 9	$X : (1,3,5b,7b,9)$	$MIDI : (0,4,6,10,14)$	
Diminished - Minor 7 - Minor 9	$X : (1,3,5b,7b,9b)$	$MIDI : (0,4,6,10,13)$	

Table A.5 Types of recorded chords with five different notes (II)

Name	Intervals	MIDI Intervals	Example
Minor - Major 7 - Major 9	$X : (1,3b,5,7,9)$	$MIDI : (0,3,7,11,14)$	
Minor - Major 7 - Minor 9	$X : (1,3b,5,7,9b)$	$MIDI : (0,3,7,11,13)$	
Minor - Minor 7 - Major 9	$X : (1,3b,5,7b,9)$	$MIDI : (0,3,7,10,14)$	
Minor - Minor 7 - Minor 9	$X : (1,3b,5,7b,9b)$	$MIDI : (0,3,7,10,13)$	
Augmented - Minor - Major 7 - Major 9	$X : (1,3b,5\#,7,9)$	$MIDI : (0,3,8,11,14)$	
Augmented - Minor - Major 7 - Minor 9	$X : (1,3b,5\#,7,9b)$	$MIDI : (0,3,8,11,13)$	
Augmented - Minor - Minor 7 - Major 9	$X : (1,3b,5\#,7b,9)$	$MIDI : (0,3,8,10,14)$	
Augmented - Minor - Minor 7 - Minor 9	$X : (1,3b,5\#,7b,9b)$	$MIDI : (0,3,8,10,13)$	
Diminished - Minor - Major 7 - Major 9	$X : (1,3b,5b,7,9)$	$MIDI : (0,3,6,11,14)$	
Diminished - Minor - Major 7 - Minor 9	$X : (1,3b,5b,7,9b)$	$MIDI : (0,3,7,11,14)$	
Diminished - Minor - Minor 7 - Major 9	$X : (1,3b,5b,7b,9)$	$MIDI : (0,3,6,10,14)$	
Diminished - Minor - Minor 7 - Minor 9	$X : (1,3b,5b,7b,9b)$	$MIDI : (0,3,6,10,13)$	

Table A.6 Types of recorded chords with six different notes (I)

Name	Intervals	MIDI Intervals	Example
Major - Major 7 - Major 9 - Perfect 11	$X : (1,3,5,7,9,11)$	$MIDI : (0,4,7,11,14,17)$	
Major - Major 7 - Minor 9 - Perfect 11	$X : (1,3,5,7,9b,11)$	$MIDI : (0,4,7,11,13,17)$	
Major - Minor 7 - Major 9 - Perfect 11	$X : (1,3,5,7b,9,11)$	$MIDI : (0,4,7,10,14,17)$	
Major - Minor 7 - Minor 9 - Perfect 11	$X : (1,3,5,7b,9b,11)$	$MIDI : (0,4,7,10,13,17)$	
Augmented - Major 7 - Major 9 - Perfect 11	$X : (1,3,5\#,7,9,11)$	$MIDI : (0,4,8,11,14,17)$	
Augmented - Major7 - Minor 9 - Perfect 11	$X : (1,3,5\#,7,9b,11)$	$MIDI : (0,4,8,11,13,17)$	
Augmented - Minor 7 - Major 9 - Perfect 11	$X : (1,3,5\#,7b,9,11)$	$MIDI : (0,4,8,10,14,17)$	
Augmented - Minor 7 - Minor 9 - Perfect 11	$X : (1,3,5\#,7b,9b,11)$	$MIDI : (0,4,8,10,13,17)$	
Diminished - Major 7 - Major 9 - Perfect 11	$X : (1,3,5b,7,9,11)$	$MIDI : (0,4,6,11,14,17)$	
Diminished - Major 7 - Minor 9 - Perfect 11	$X : (1,3,5b,7,9b,11)$	$MIDI : (0,4,6,11,13,17)$	
Diminished - Minor 7 - Major 9 - Perfect 11	$X : (1,3,5b,7b,9,11)$	$MIDI : (0,4,6,10,14,17)$	
Diminished - Minor 7 - Minor 9 - Perfect 11	$X : (1,3,5b,7b,9b,11)$	$MIDI : (0,4,6,10,13,17)$	

Table A.7 Types of recorded chords with six different notes (II)

Name	Intervals	MIDI Intervals	Example
Minor - Major 7 - Major 9 - Perfect 11	$X : (1, 3b, 5, 7, 9, 11)$	$MIDI : (0, 3, 7, 11, 14, 17)$	
Minor - Major 7 - Minor 9 - Perfect 11	$X : (1, 3b, 5, 7, 9b, 11)$	$MIDI : (0, 3, 7, 11, 13, 17)$	
Minor - Minor 7 - Major 9 - Perfect 11	$X \cdot (1, 3b, 5, 7b, 9, 11)$	$MIDI : (0, 3, 7, 10, 14, 17)$	
Minor - Minor 7 - Minor 9 - Perfect 11	$X \cdot (1, 3b, 5, 7b, 9b, 11)$	$MIDI : (0, 3, 7, 10, 13, 17)$	
Augmented - Minor - Major 7 - Major 9 - Perfect 11	$X : (1, 3b, 5\#, 7, 9, 11)$	$MIDI : (0, 3, 8, 11, 14, 17)$	
Augmented - Minor - Major 7 - Minor 9 - Perfect 11	$X : (1, 3b, 5\#, 7, 9b, 11)$	$MIDI : (0, 3, 8, 11, 13, 17)$	
Augmented - Minor - Minor 7 - Major 9 - Perfect 11	$X : (1, 3b, 5\#, 7b, 9, 11)$	$MIDI : (0, 3, 8, 10, 14, 17)$	
Augmented - Minor - Minor 7 - Minor 9 - Perfect 11	$X : (1, 3b, 5\#, 7b, 9b, 11)$	$MIDI : (0, 3, 8, 10, 13, 17)$	
Diminished - Minor - Major 7 - Major 9 - Perfect 11	$X : (1, 3b, 5b, 7, 9, 11)$	$MIDI : (0, 3, 6, 11, 14, 17)$	
Diminished - Minor - Major 7 - Minor 9 - Perfect 11	$X : (1, 3b, 5b, 7, 9b, 11)$	$MIDI : (0, 3, 6, 11, 13, 17)$	
Diminished - Minor - Minor 7 - Major 9 - Perfect 11	$X : (1, 3b, 5b, 7b, 9, 11)$	$MIDI : (0, 3, 6, 10, 14, 17)$	
Diminished - Minor - Minor 7 - Minor 9 - Perfect 11	$X : (1, 3b, 5b, 7b, 9b, 11)$	$MIDI : (0, 3, 6, 10, 13, 17)$	

Table A.8 Types of recorded chords with seven different notes (I)

Name	Intervals	MIDI Intervals	Example
Major - Major 7 - Major 9 - Perfect 11 - Major 13	$X : (1,3,5,7,9,11,13)$	$MIDI : (0,4,7,11,14,17,21)$	
Major - Major 7 - Minor 9 - Perfect 11 - Major 13	$X : (1,3,5,7,9b,11,13)$	$MIDI : (0,4,7,11,13,17,21)$	
Major - Minor 7 - Major 9 - Perfect 11 - Major 13	$X : (1,3,5,7b,9,11,13)$	$MIDI : (0,4,7,10,14,17,21)$	
Major - Minor 7 - Minor 9 - Perfect 11 - Major 13	$X : (1,3,5,7b,9b,11,13)$	$MIDI : (0,4,7,10,13,17,21)$	
Augmented - Major 7 - Major 9 - Perfect 11 - Major 13	$X : (1,3,5\#,7,9,11,13)$	$MIDI : (0,4,8,11,14,17,21)$	
Augmented - Major 7 - Minor 9 - Perfect 11 - Major 13	$X : (1,3,5\#,7,9b,11,13)$	$MIDI : (0,4,8,11,13,17,21)$	
Augmented - Minor 7 - Major 9 - Perfect 11 - Major 13	$X : (1,3,5\#,7b,9,11,13)$	$MIDI : (0,4,8,10,14,17,21)$	
Augmented - Minor 7 - Minor 9 - Perfect 11 - Major 13	$X : (1,3,5\#,7b,9b,11,13)$	$MIDI : (0,4,8,10,13,17,21)$	
Diminished - Major 7 - Major 9 - Perfect 11 - Major 13	$X : (1,3,5b,7,9,11,13)$	$MIDI : (0,4,6,11,14,17,21)$	
Diminished - Major 7 - Minor 9 - Perfect 11 - Major 13	$X : (1,3,5b,7,9b,11,13)$	$MIDI : (0,4,6,11,13,17,21)$	
Diminished - Minor 7 - Major 9 - Perfect 11 - Major 13	$X : (1,3,5b,7b,9,11,13)$	$MIDI : (0,4,6,10,14,17,21)$	
Diminished - Minor 7 - Minor 9 - Perfect 11 - Major 13	$X : (1,3,5b,7b,9b,11,13)$	$MIDI : (0,4,6,10,13,17,21)$	

Table A.9 Types of recorded chords with seven different notes (II)

Name	Intervals	MIDI Intervals	Example
Minor - Major 7 - Major 9 - Perfect 11 - Major 13	$X : (1, 3b, 5, 7, 9, 11, 13)$	$MIDI : (0, 3, 7, 11, 14, 17, 21)$	
Minor - Major 7 - Minor 9 - Perfect 11 - Major 13	$X : (1, 3b, 5, 7, 9b, 11, 13)$	$MIDI : (0, 3, 7, 11, 13, 17, 21)$	
Minor - Minor 7 - Major 9 - Perfect 11 - Major 13	$X : (1, 3b, 5, 7b, 9, 11, 13)$	$MIDI : (0, 3, 7, 10, 14, 17, 21)$	
Minor - Minor 7 - Minor 9 - Perfect 11 - Major 13	$X : (1, 3b, 5, 7b, 9b, 11, 13)$	$MIDI : (0, 3, 7, 10, 13, 17, 21)$	
Augmented - Minor - Major 7 - Major 9 - Perfect 11 - Major 13	$X : (1, 3b, 5\#, 7, 9, 11, 13)$	$MIDI : (0, 3, 8, 11, 14, 17, 21)$	
Augmented - Minor - Major 7 - Minor 9 - Perfect 11 - Major 13	$X : (1, 3b, 5\#, 7, 9b, 11, 13)$	$MIDI : (0, 3, 8, 11, 13, 17, 21)$	
Augmented - Minor - Minor 7 - Major 9 - Perfect 11 - Major 13	$X : (1, 3b, 5\#, 7b, 9, 11, 13)$	$MIDI : (0, 3, 8, 10, 14, 17, 21)$	
Augmented - Minor - Minor 7 - Minor 9 - Perfect 11 - Major 13	$X : (1, 3b, 5\#, 7b, 9b, 11, 13)$	$MIDI : (0, 3, 8, 10, 13, 17, 21)$	
Diminished - Minor - Major 7 - Major 9 - Perfect 11 - Major 13	$X : (1, 3b, 5b, 7, 9, 11, 13)$	$MIDI : (0, 3, 6, 11, 14, 17, 21)$	
Diminished - Minor - Major 7 - Minor 9 - Perfect 11 - Major 13	$X : (1, 3b, 5b, 7, 9b, 11, 13)$	$MIDI : (0, 3, 6, 11, 13, 17, 21)$	
Diminished - Minor - Minor 7 - Major 9 - Perfect 11 - Major 13	$X : (1, 3b, 5b, 7b, 9, 11, 13)$	$MIDI : (0, 3, 6, 10, 14, 17, 21)$	
Diminished - Minor - Minor 7 - Minor 9 - Perfect 11 - Major 13	$X : (1, 3b, 5b, 7b, 9b, 11, 13)$	$MIDI : (0, 3, 6, 10, 13, 17, 21)$	

References

1. A.P. Klapuri,"Multiple fundamental frequency estimation based on harmonicity and spectral smoothness", IEEE Transaction on Speech and Audio Processing, vol.11, no.6, pp. 804–816, 2003.
2. M. Marolt, "A connectionist approach to automatic transcription of polyphonic piano music", IEEE Transaction on Multimedia, vol. 6, no.3, pp. 439–449, 2004.
3. I.P. Bello, L. Daudet and M.B. Sandler, "Automatic piano transcription using frequency and time-domain information", IEEE Transaction on Audio, Speech and Language Processing, vol. 14, pp. 2242–2251, Nov. 2006.
4. G.E. Poliner and D.P.W. Ellis, "A discriminative model for polyphonic piano transcription", EURASIP Journal on Advances in Signal Processing, vol. 8, pp. 1–9, 2007.
5. I. Barbancho, A.M. Barbancho, A. Jurado and L.J. Tardón, "Transcription of Piano Recordings", Applied Acoustics, vol. 65, pp. 1261–1287, December 2004.
6. A.M. Barbancho, L.J. Tardón and I. Barbancho, "PIC Detector por Piano Chords", EURASIP Journal on Advances in Signal Processing, pp.1–12, 2010.
7. A.M. Barbancho, I. Barbancho, B. Soto and L.J. Tardón, "SIC Receiver for Polyphonic Piano Music", in IEEE International Conference on Acoustics, Speech and Signal Processing (ICASSP 2011), pp. 377–380, May 2011.
8. A.M. Barbancho, I. Barbancho, J. Alamos and L.J. Tardón, "Polyphony number estimator for piano recordings using different spectral patterns", in Audio Engineering Society Convention (AES 128th), 2009.
9. M. Goto, "Development of the RWC Music Database", Proceedings of the 18th International Congress on Acoustics, April 2004.
10. "Grove Music Online: the world's premier authority on all aspects in music". Oxford University Press. www.oxfordmusiconline.com
11. F. Opolko and J. Wapnick, "McGill University Master Samples. A 3-DVD Set", 2006.
12. "The Complete MIDI 1.0 Detailed Specification", 2nd ed., The MIDI Manufacturers Association, 1996, website: www.midi.org.
13. W. Apel, "Harvard Dictionary of Music", 2nd. ed., The Belknap Press of Harvard University Press, Cambridge, Massachusetts 2000.
14. C. L. Krumhansl, "Cognitive Foundation of Musical Pitch",Oxford University Press, New York, NY, USA, 1990.
15. www.music-ir.org/mirex/
16. A. Klapuri and M. Davy, "Signal processing methods for music transcription", Springer, 2006.
17. J. M. Barbour, "Tuning and Temperament: A Historical Survey", Dover Publications 2004.

A.M. Barbancho et al., *Database of Piano Chords: An Engineering View of Harmony*,   47
SpringerBriefs in Electrical and Computer Engineering,
DOI 10.1007/978-1-4614-7476-0, © The Author(s) 2013

18. P. Vos and B. G. V. Vianen, "Thresholds for discrimination between pure and tempered intervals: The relevance of nearly coinciding harmonics", Journal of the Acoustical Society of America, vol. 77, pp.176–187, 1984.
19. I. Kosuke, M. Ken'Ichi and N. Tsutomu, "Ear advantage and consonance of dichotic pitch intervals in absolute-pitch possessors. Brain and cognition", vol. 53, no.3, pp.464–471, 2003.
20. C. de la Bandera, S. Sammartino, I. Barbancho and L.J. Tardón, "Evaluation of music similarity based on tonal behavior", In 7th International Symposium On Computer Music Modeling and Retrieval Málaga (Spain), June 21-24, 2010.
21. K. Jensen, "Envelope model of isolated musical sounds", In Proceeedings of the 2nd COST G-6 Workshop on Digital Audio Effects (DAFx99), NTNU, Trondheim, December 9-11, 1999.
22. H.L.F. von Helmholtz, "On the Sensations of Tone as a Physiological Basis for the Theory of Music", 4th edition. Trans., A.J. Ellis, New York: Dover, 1954.
23. T.D. Rossing, F.R. Moore and P.A. Wheeler, "The Science of Sound", 3rd edition, Addison Wesley, 2002.
24. P.D. Lehrman and T. Tully, "MIDI for the Professional", Amsco Publications, 1993.
25. A. V. Oppenheim and R.W. Schafer, "Discrete-Time Signal Processing", Prentice Hall, 1989.
26. R. Cruz, A. Ortiz, A.M Barbancho and I. Barbancho, "Unsupervised classification of audio signals by self-organizing maps and bayesian labeling". International Conference on Hybrid Artificial Intelligence Systems. 2012. LNAI 7208.
27. A. Ortiz, L. Tardón, A.M. Barbancho and I. Barbancho "Unsupervised and neural hybrid techniques for audio signal classification". Independent Component Analysis for Audio and Biosignal applications. In-Tech, Viena, Austria, 2012, ISBN: 980-953-307-197-3.

Index

A.M. Barbancho et al., *Database of Piano Chords: An Engineering View of Harmony*,   49
SpringerBriefs in Electrical and Computer Engineering,
DOI 10.1007/978-1-4614-7476-0, © The Author(s) 2013